YOUR KNOWLEDGE HAS VALUE

- We will publish your bachelor's and
  master's thesis, essays and papers

- Your own eBook and book -
  sold worldwide in all relevant shops

- Earn money with each sale

Upload your text at www.GRIN.com
and publish for free

Markus Meyer

Surface sealing and the water balance

Germany and Belgium

GRIN Verlag

Bibliografische Information der Deutschen Nationalbibliothek:

Die Deutsche Bibliothek verzeichnet diese Publikation in der Deutschen National-
bibliografie; detaillierte bibliografische Daten sind im Internet über http://dnb.d-
nb.de/ abrufbar.

Imprint:

Copyright © 2011 GRIN Verlag GmbH
Druck und Bindung: Books on Demand GmbH, Norderstedt Germany
ISBN: 978-3-656-09532-3

This book at GRIN:

http://www.grin.com/en/e-book/184670/surface-sealing-and-the-water-balance

Surface sealing and the water balance

Germany and Belgium

Markus Meyer

Course:
MX0105 Water Resource Dilemmas, Uncertainty and Complexity: The Biophysical Basis

04.11.2011

Abstract: This report investigates the relationship between the degree of imperviousness – surface sealing – and the water balance – surface runoff, evapotranspiration and infiltration. It synthesizes case studies for the cities Leipzig, Dessau and Munich in Germany and the Grote-Nete catchment in Belgium and compares the effect of different urban land uses on the water balance. The annual precipitation ranges from 530 to 950 mm. Depending on the respective hydrological properties and land use characteristics a linear increase of surface runoff with increasing degree of imperviousness could be found for all of the German studies. Evapotranspiration and infiltration decline with an increasing degree of imperviousness. The relationship for both is not distinct as for surface runoff. The impact on the water balance does not significantly deviate for different levels of precipitation.

1. Introduction

Surface sealing or soil surface sealing is the packing or compaction of the earth surface to hinder the infiltration of fluids (EEA, 2011). A measure to compare different types of land use is to set the amount of impervious or totally sealed area in relation to the total area considered. This value is also named the degree of imperviousness. (Leopold, 1968)

In Europe, 1.81 percent of the total area is sealed. This amount seems rather low and does not seem to affect humans or the environment. (EEA, 2010) But in countries such as Malta, Belgium, the Netherlands or Germany it reaches beyond five percent of the total area as shown in the figure 1.

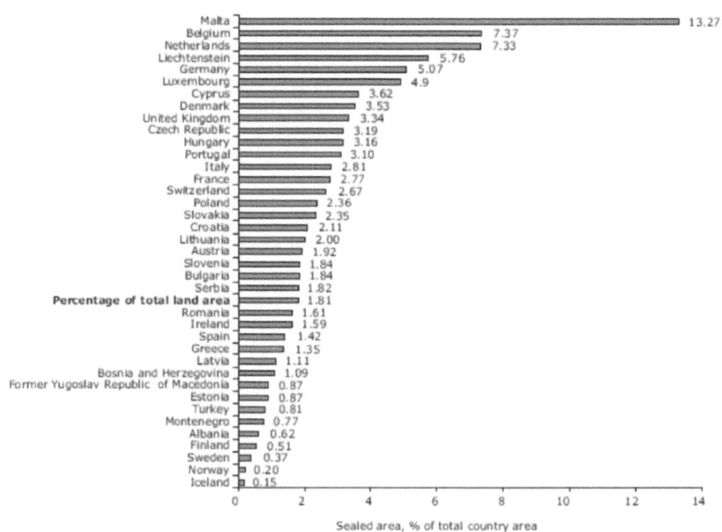

Figure 1: Degree of imperviousness in European countries (2006) (EEA, 2010)

It is known that the major problems of surface sealing occur in urban areas with higher degrees of imperviousness, which ranges e.g. for European capitals from 20 percent in Stockholm to about 80 percent in Bucharest and this is shown in figure 2. Additionally, the average annual rate of land take was for the period 2000 to 2006 111,788 ha per year and per country or equal to 0.6 percent of the total area of each country (36 European countries) and increased by nine percent compared with the period 1990-2000 (24 European countries) (EEA, 2011). Assuming that that this trend will persist, the degree   of   imperviousness   indicated   in   figure   2   will   increase   further.

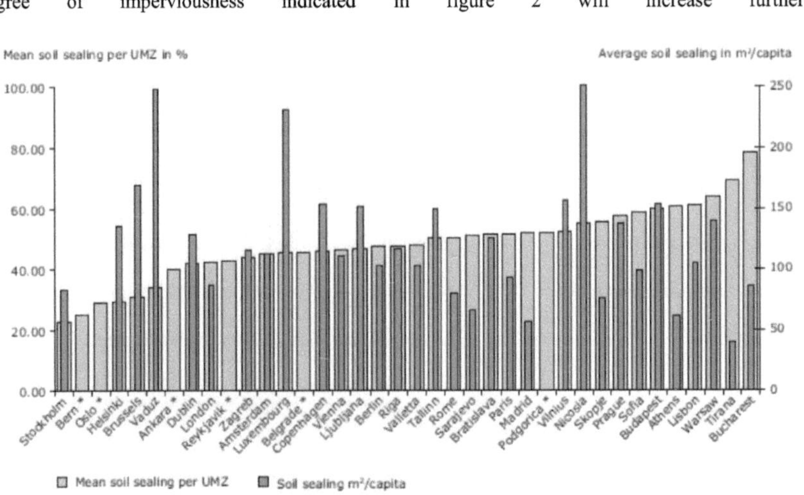

Figure 2: Average surface sealing in % of Urban Morphological Zone and per inhabitant (EEA, 2010)

The trade-offs from surface sealing mainly occur in urbanized areas as they have a higher degree of imperviousness as paved roads or buildings are built much more densely. Especially a decrease of the permeability of the soil increases surface runoff and decreases infiltration. Due to the lower infiltration to recharge the groundwater, groundwater aquifers are not sufficiently recharged anymore and the water table declines (Haase & Nuissl, 2007). The filtering capacity of the soil is only used to a much smaller extent as less water moves through the soil. (Emmerling & Udelhoven, 2002) Due to soil surface sealing, the water storage capacity of the soil - a buffer for the variability of precipitation – can be used less due to hindered infiltration. As a consequence the amplitude between peak and base flow increases and the discharge occurs at different times – more at rainstorm events, less in between. The base flow of receiving surface water bodies such as rivers decreases. This means that the water provision in dry periods is weakened as the soil water cannot contribute to the surface runoff the more the soil is sealed. At peak flows the discharge rate and velocity of surface runoff is increased compared to unsealed soils as sealed soils respond faster to rainstorm events. (Pauleit & Duhme, 2000)(Rodhe &

Kilingtveit, 1997) A higher share of local precipitation reaches rivers and finally the ocean and is not locally available anymore (Scalenghe & Marsan, 2009).

Also the atmospheric water content in sealed urban areas is lower than in rural areas due to the lack of vegetation, which significantly increases evapotranspiration due to interception. Especially forests with a high-leaf-area index hinder more water to reach the ground so that surface runoff is reduced respectively in urban areas increased. (Hundecha & Bárdossy, 2004) The locally decreased evapotranspiration alters the microclimate in cities and produces a negative feedback. The humidity of the air will decrease. As a consequence, the precipitation in the cities and in areas, where the humidity is transported to, will also decline so that overall amount of water in all components of the water balance will decline (Göbel, Starke, Meßer, & Coldewey, 2011)

These major impacts are also displayed in the generic figure 3.

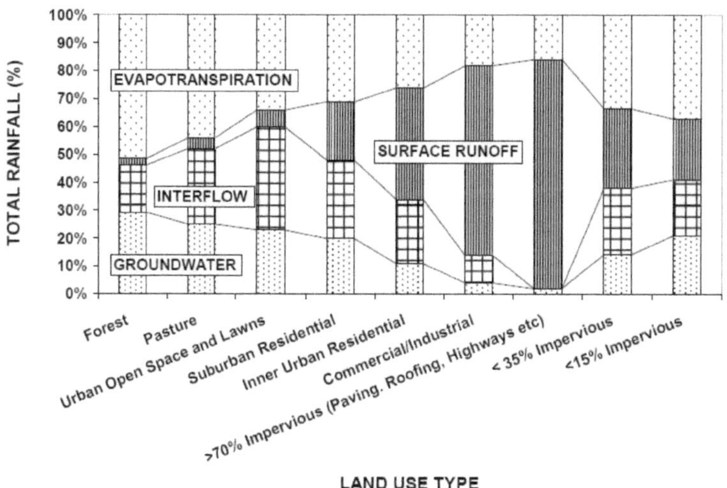

Figure 3: Urban water balance for different land use categories and degrees of imperviousness (Ellis, 2008)

Additionally, it has to be considered that urban freshwater – unequal to rural – demand can only be met by additional water supply from their surroundings or by extraction of groundwater *(EEA, 2010)*. So the groundwater resources are threatened from the supply and demand side.

The focus of this report is to find empirical evidence in different studies in urban environments for the above named negative effects of surface sealing on the water balance. Especially the relationship between the components surface runoff, infiltration as well as evapotranspiration and the degree of imperviousness should be investigated at different levels of precipitation. The focus is set on Germany as the total amount of sealed surface in the European countries is highly concentrated, although it is one of the larger countries of Europe. So it can be expected that both more studies on surface sealing and the water balance are available due to the larger size than in Malta or the Netherlands and also most of negative effects of a high degree of surface sealing will be found. An additional case study will be taken from Belgium to test, whether the findings for Germany are also valid for other European countries. In the water balance, the water quantity and not the water quality such as the filtering capacity of the soil is investigated.

1.1 Areas of study

	Leipzig (Haase & Nuissl, 2007)	Dessau (Riemann, 1999)	Munich (Pauleit & Duhme, 2000)	Grote-Nete (Belgium) (Asefa, Wang, Batelaan, & De Smedt, 1999)
Size in km²	62.72	30	50	525.5
Average precipitation (mm/year)	531-596	530	950	743-800
Degree of imperviousness	53 %	NN	42 %	7%
Catchment	Weiße Elster	Elbe	Isar	Grote-Nete

Table 1: Overview of areas of study

The areas of study vary from a size of 30 km² in Dessau to 525.5 km² for the catchment of the river Grote-Nete in Belgium. The studies on the water balance in the two eastern German cities Leipzig and Dessau and the southern German city Munich are conducted in urbanized areas. The Belgian study focuses on a catchment with rural and urban land use. This can also be seen by the higher degrees of imperviousness for the whole area in Leipzig and Munich compared with the low degree of impervious of the Belgian Grote-Nete catchment. The respective catchment is indicated in the last row. The spatial distribution of the studies can be seen in figure 4:

Figure 4: Map of the location of studies (Google Inc., 2011)

We also have a range of average annual precipitation values from 530 to 950 mm per year so that the effect of the level of precipitation on the water balance at different degrees of imperviousness can be studied.

2. Methods

		Leipzig (Haase & Nuissl, 2007)	Dessau (Riemann, 1999)	Munich (Pauleit & Duhme, 2000)	Grote-Nete (Belgium) (Asefa, Wang, Batelaan, & De Smedt, 1999)
Model name		ABIMO/ Messer's model	NN	NN	WetSpass
Approach for water balance calculation		Grid cell	NN	Grid cell	Grid cell
Input data	Climate data (annual means)	Precipitation , potential evapotranspiration (gridded)	Precipitation (homogenous for the whole area)	Precipitation (homogenous for the whole area)	Precipitation, potential evapotranspiration, wind speed (gridded)
	Catchment characteristics	Land use, slope (40mx40m), groundwater level depth	NN	Land use	Land use (50m x 50m), slope, groundwater level depth
	Soil properties	Grain size, field capacity	NN	none	Hydraulic properties, empirical coefficients for evapotranspiration and surface runoff modelling
Modelling period		2003	NN	1991	1999

Table 2: Overview of data input for the water balance modeling for the areas of study

2.1 Calculation of the water balance

For the city of Leipzig, the ABIMO model estimates the surface runoff from the mean actual evapotranspiration and precipitation. The mean actual evapotranspiration is estimated from the potential evapotranspiration and precipitation by the BAGROV-relation(Bagrov, 1953), which uses soil properties and land use data. Additionally, it is assumed that with decreasing precipitation actual evapotranspiration also declines. Using the estimated actual evapotranspiration, the precipitation data and the catchment characteristics, the surface runoff is estimated with Messer's model(Messer, 1997). The underlying idea is that surface runoff increases with increasing slope and decreasing depth of groundwater table and especially increasing degree of imperviousness (Haase & Nuissl, 2007)

For the model used in Dessau no further methodological explanations are given. The model in Munich is based on the above indicated data categories, the coefficients to estimate the net infiltration – groundwater recharge without interflow – and surface runoff are taken from other studies for other areas in Germany and land use is the only defining parameter for differences in the water balance. Evapotranspiration is assumed to be the residual of precipitation substracting surface runoff and net infiltration. (Pauleit & Duhme, 2000)

For the catchment Grote-Nete in Belgium, the model Water and Energy Transfer between Soil, Plants and Atmosphere under Steady State (WetSpass) was used by Asefa, Wang, Batelaan and De Smedt (1999) to estimate water balance for different land use categories. The water balance is calculated with the above named input data for grid cells and is based on values for a vegetated and a non-vegetated part. For the vegetated part, actual evapotranspiration, interception by vegetation, surface runoff and infiltration for a river basin are calculated. For the non-vegetated part, the interception by vegetation is not estimated.

The actual evapotranspiration is calculated from the potential evapotranspiration with a coefficient considering the sand content in the soil and the available water in the soil and from precipitation. Interception is assumed to be 15 percent of precipitation and included in the evapotranspiration value. The surface runoff is estimated by using the coefficient of Mallants and Feyen(Mallants & Feyen, 1990) for vegetated and non-vegetated surface as a share of precipitation less actual evapotranspiration – and for vegetated surfaces also interception.

To sum up, precipitation P, evapotranspiration ET, surface runoff Q and the amount available for infiltration I are modeled with parts or all of the data shown in figure 5. The indicated evapotranspiration calculated for Munich might be overestimating the actual value as the net value of groundwater recharge is modeled so that the interflow is included in the value for evapotranspiration (Pauleit & Duhme, 2000).

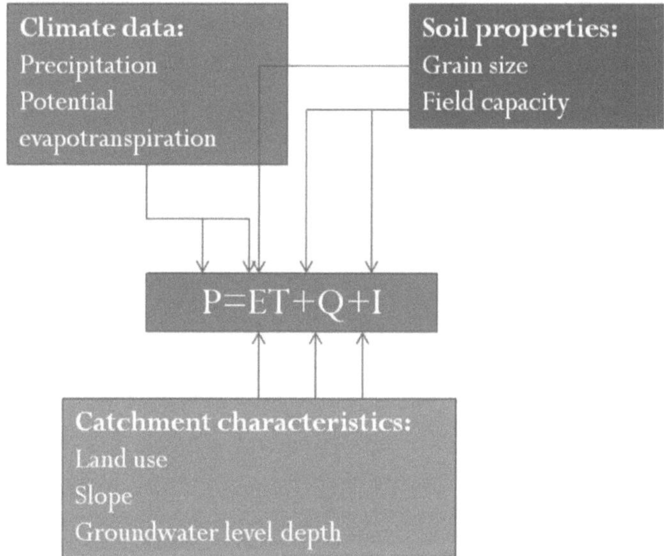

Figure 5: Overview of the data input for calculating the water balance

2.2 Matching of different studies by degree of imperviousness

The water balance was set up for the components actual evapotranspiration, surface runoff and groundwater recharge. The comparison of different values is done based on the degrees of imperviousness of urban soil surfaces. In Leipzig, the water balance is calculated for the five bands of imperviousness shown in table 4 – no value is calculated for totally pervious surfaces (degree of imperviousness: zero percent).

Categories of impervious land (%)	Land-use categories
0	Agriculture/forestry, raw materials extraction, areas of water
>0–20	Green spaces, parks, gardens, cemeteries
>20–40	Railway tracks, sports facilities, health/social services, ribbon development
>40–60	Large housing estates (flats and houses), utilities/waste disposal, military
>60–80	Older villages, 1990s housing estates, education/research
>80–100	Centre, commercial space, the service sector, trade show venues, roads

Table 3: Degrees of impervious and related land use classes for Leipzig (Germany) (Haase & Nuissl, 2007)

For the study of Dessau, the water balance was indicated for six different degrees of imperviousness over the whole range without naming the different land use categories (Riemann, 1999). For Munich the water balance was only indicated for two narrowly defined land use classes and the average degree of imperviousness of 42.3 percent. The two classes are detached houses and densely built-up blocks with a degree of imperviousness of 30.5 percent respectively 79.9 percent shown in table 4.

S. No.	Land cover types	No. of units	Area (ha)	Percentage of study area (%)	Sealed surfaces (%)[a]		Built-up (%)[a]		Vegetation (%)[b]	
					Mean	S.D.	Mean	S.D.	Mean	S.D.
1	Detached houses	83	724.99	14.78	30.5	15.3	15.3	6.9	66.8	14.8
2	Terraced houses	25	107.42	2.19	46.4	14.9	29.0	11.5	52.5	15.4
3	Multistory houses	219	836.27	17.04	51.5	17.3	29.5	11.0	43.8	17.5
4	Multistory blocks	54	513.39	10.46	79.9	18.1	44.3	10.5	19.1	16.9
5	Factory buildings	43	129.13	2.63	46.3	27.2	23.9	16.1	31.8	23.2
6	Multistory/factory buildings	56	432.90	8.82	75.7	17.7	35.2	14.6	12.8	14.0
7	Mixed use	11	38.36	0.78	36.7	13.8	22.0	9.7	59.1	11.1
8	Special buildings	4	24.96	0.51	50.0	20.4	22.5	6.5	41.3	21.4
9	Construction sites	28	122.07	2.49	8.0	20.4	3.9	8.6	16.0	17.4
10	Large car parks	9	31.90	0.65	59.3	36.1	5.8	11.3	15.9	10.2
11	Roads	76	299.62	6.11	90.8	10.4	0.2	1.0	6.9	8.8
12	Railways	10	261.32	5.33	7.4	10.4	1.2	2.3	8.1	8.1
13	Lakes and ponds	7	35.82	0.73	0.0	0.0	0.0	0.0	0.0	0.0
14	Streams	23	29.99	0.61	9.3	14.9	0.3	1.2	52.8	28.5
15	Woodlands	24	226.30	4.61	0.5	1.4	0.0	0.2	96.7	3.1
16	Hedgerows/woodlots	46	47.02	0.96	13.3	24.8	0.7	2.0	83.4	24.3
17	Parks and green spaces	100	357.72	7.29	12.7	19.8	0.9	3.7	77.6	21.4
18	Cemeteries	3	55.85	1.14	2.3	2.1	1.3	1.5	88.0	9.8
19	Allotment gardens	35	108.19	2.20	16.1	4.7	11.0	3.8	77.2	7.6
20	Sports fields	30	162.64	3.31	14.1	17.1	3.7	6.4	73.6	18.2
21	Meadows and pastures	18	78.93	1.61	7.7	17.1	4.1	16.5	88.1	20.8
22	Extensive grasslands	55	88.33	1.80	1.8	3.7	0.3	1.1	80.1	20.0
23	Arable fields	37	134.21	2.74	1.2	2.8	0.2	0.8	96.7	4.0
24	Horticulture/nurseries	32	59.19	1.21	25.1	23.4	19.9	20.7	68.8	24.6
	Totals/average[d]	1028	4906.52	100.00	42.3	–	18.2	–	45.0	–

[a] Percentage of unit area.

[b] Percentage of annual precipitation of 950 mm. Evapotranspiration is already subtracted from infiltration.

[c] Runoff from a 40 mm rainstorm per hour.

[d] The average values for the whole study area of 4907 ha are given here instead of the (unweighted) mean values

Table 4: Surface sealing of different land use classes in Munich (Germany) (Pauleit & Duhme, 2000)

For the catchment Grote-Nete, the water balance is only calculated for the following land use categories without indicating the degree of imperviousness:

Land use
Crop/mixed farming
Short grass
Evergreen pine forest
D. broad leaf forest
Deciduous shrubs
Non-vegetated
Open water

Table 5: Land use classes in Grote-Nete (Belgium) (Asefa, Wang, Batelaan, & De Smedt, 1999)

The first five and the last land use classes could be attributed to the degree of imperviousness of zero as indicated in the user manual for the WetSpass model. For the non-vegetated land use class no degree of imperviousness is indicated so that the values are not usable. But further data for the land use class *impervious* representing urban land use with a degree of imperviousness of 70 percent indicated in the user manual was available (De Smedt & Liu, 2004). The obtained average degree of imperviousness with the attributed values for the different degrees of imperviousness of 7 percent shown in table 1 seems reasonable as it matches with the Belgian average degree of imperviousness of about 7.37 percent shown in figure 2.

2.3 Estimation of the relationship of the water balance and degree of imperviousness

According to Bhaduri et al. (2001) and Assouline and Mualem (2002) the increase of surface runoff is linearly dependent on the degree of imperviousness. Whether this relationship is also holding for Leipzig, Dessau and Munich will be tested by regression analysis. In a study for the city of Münster in Germany a linear dependency of evapotranspiration on the degree of imperviousness was discovered (Göbel, Starke, Meßer, & Coldewey, 2011). This will also be tested for Leipzig, Munich and Dessau. Additional, regressions will be effected for infiltration for Leipzig, Munich and Dessau to investigate the relationship with the degree of imperviousness. For Grote-Nete, this is not tested as only two data points are given.

To test whether the dependency expressed by the coefficient of determination is significant a one-sided t-test is used (Rasch, Friese, Hofmann, & Neumann, 2010). It is tested against the zero-hypothesis that there is no correlation between the variables with the correlation coefficient r equal to zero (Rasch, Friese, Hofmann, & Neumann, 2010). The test is effected at a significance level of 99.5%.

3. Results

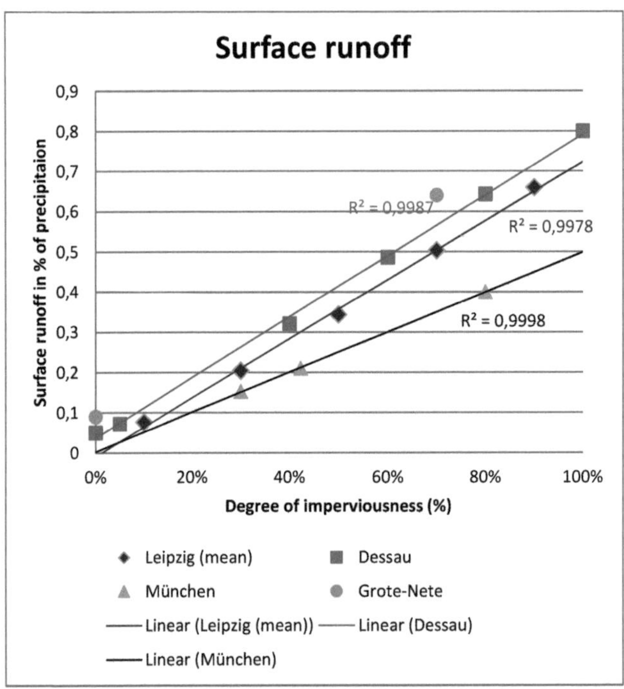

Figure 6: Surface runoff and degree of imperviousness[1]

We can see in figure 6 a high increase of surface runoff between 0 to 10 percent of total precipitation to 50 to 80 percent from 0 to 100 percent impervious surfaces. The surface runoff shows for Leipzig, Dessau and Munich a linear relationship between degree of imperviousness and surface runoff with a

---

[1] Calculations based on data from (Asefa, Wang, Batelaan, & De Smedt, 1999), (Haase & Nuissl, 2007), (Riemann, 1999) and (Pauleit & Duhme, 2000).

high coefficient of determination. The slope of the increase of surface runoff with increasing degree of imperviousness in Munich is lower than for Dessau and Leipzig.

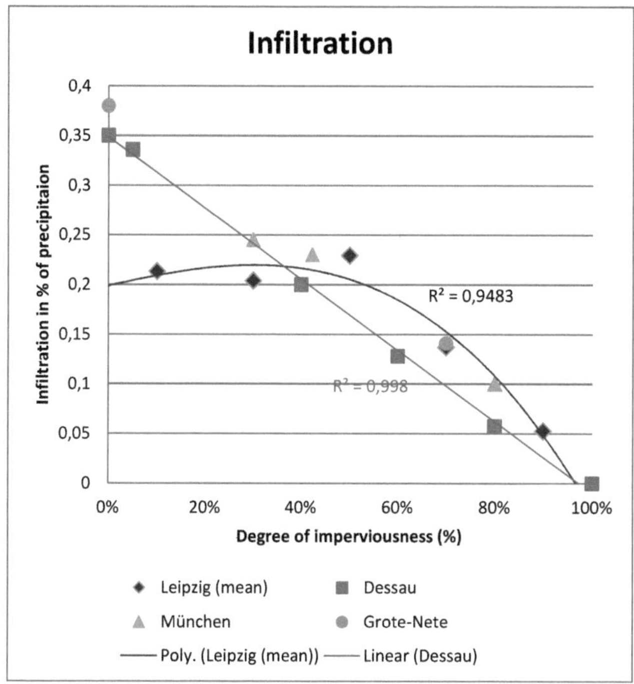

Figure 7: Infiltration and degree of imperviousness[2]

We can see in figure 7 and 8 that the infiltration in Munich, Dessau and Grote-Nete decreases from a level of 35 to 40 percent of total precipitation for unsealed soils to 5 to 15 percent for areas impervious to 80 percent. For Dessau a linear relationship between degree of imperviousness and infiltration with the highest coefficient of determination could be identified. This is not given for the other studies as also shown in the graph below with more data points for Munich, where an exponential relationship seems stronger than a linear one. For Leipzig the infiltration seems to stay constant between 20 and 23 percent of total precipitation until a degree of imperviousness between 50 and 70 percent and only

---

[2] Calculations based on data from (Asefa, Wang, Batelaan, & De Smedt, 1999), (Haase & Nuissl, 2007), (Riemann, 1999) and (Pauleit & Duhme, 2000).

declines with a further increasing degree of imperviousness. The strongest modeling seems to be poly-

nomial of 2nd degree.

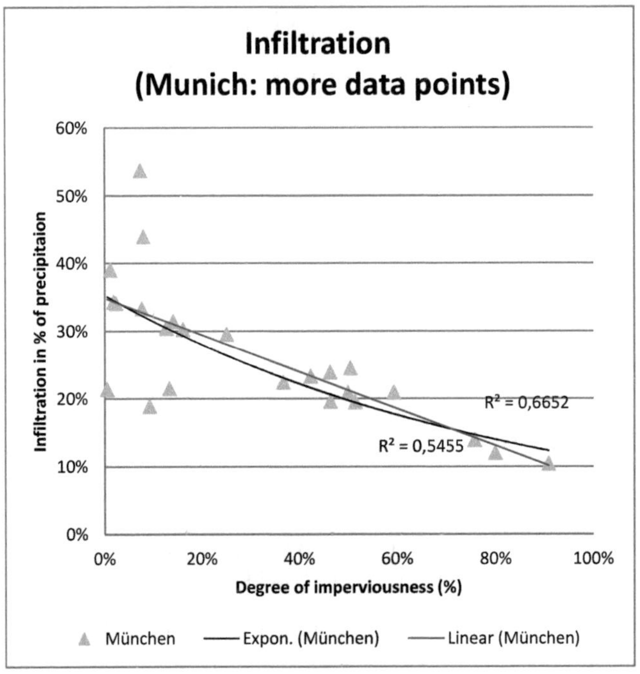

Figure 8: Infiltration and degree of imperviousness in Munich[3]

---

[3] Calculations based on data from (Pauleit & Duhme, 2000).

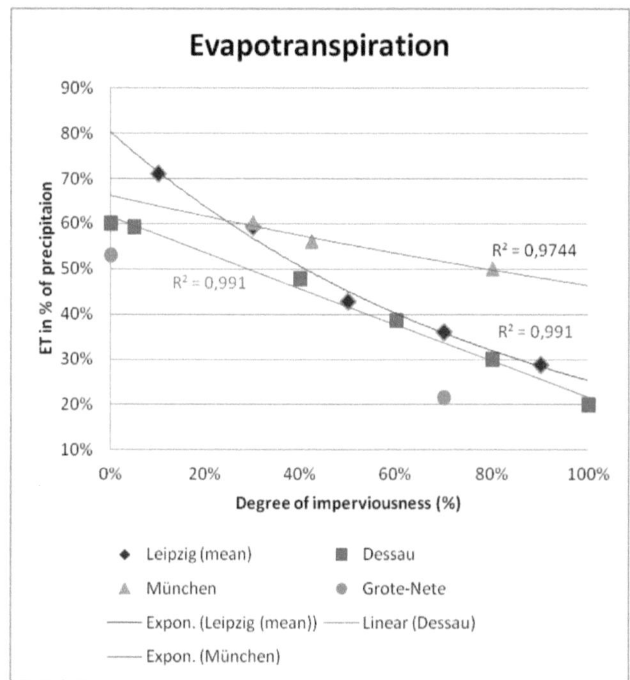

Figure 9: Evapotranspiration and degree of imperviousness[4]

We could see in figure 9 that evapotranspiration in Munich, Dessau and Grote-Nete decreases from a level of 50 to 70 percent of total precipitation for unsealed soils to 20 to 50 percent for areas impervious to 80 percent. For Dessau a linear relationship and for Leipzig and Munich an exponential relationship between degree of imperviousness and evapotranspiration have the highest coefficient of determination.

Formula	Study	Evapotranspiration	Infiltration	Surface runoff
t-t(0.5%;Df.)	Leipzig	20.08	4.82	46.55
t-t(0.5%;Df.)	Dessau	20.08	49.11	62.28
t-t(0.5%;Df.)	Munich	-5.39	5.80	85.90

Table 6: Significance test of correlation between degree of imperviousness and water balance components

---

[4] Calculations based on data from (Asefa, Wang, Batelaan, & De Smedt, 1999), (Haase & Nuissl, 2007), (Riemann, 1999) and (Pauleit & Duhme, 2000).

Table 6 shows t values for the different coefficients of determination of the regression lines minus the t value with a probability of 99.5 percent at the respective degrees of freedom. We see that only the determined relationship for evapotranspiration and degree of imperviousness for Munich is not significant. It is significant with a probability of 95 percent with t-t(5 percent;Df.) equal to 4.39.

In figure 10, it can be spatially seen for Leipzig comparing the years 1870 and 2003 that an increase in urban land use with a higher degree of imperviousness is accompanied by an significantly increase of surface runoff so that the flooding risk is increased as well.

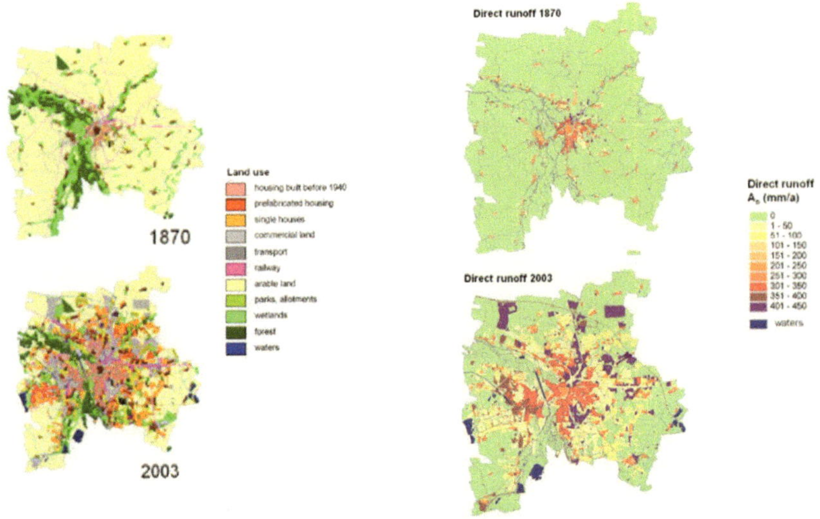

Figure 10: Land use and runoff map for Leipzig 1870 and 2003 (Haase, 2009)

4. Discussion

Even if the methodology and the different parameters defining the water balance are locally specific, the increase of surface runoff and the decline of infiltration and evapotranspiration with increasing degree of imperviousness can be seen in all studies. It is also necessary to consider that surface sealing is only one parameter of human action. Nevertheless, it severely disturbs the water balance. This could be seen for different levels of precipitation from 530 mm to 950 mm per year.

One consequence from increased runoff is an increased risk of flooding. As the degree of imperviousness indicates, how strongly an area is urbanized, it can be said that increasing urbanization increases the risk of flooding. It cannot not be generally said that the slope of runoff increase with increasing degree of imperviousness is different for different levels of precipitation. For a higher level of precipitation in Munich, the share of surface runoff from total precipitation is lower . But in the Grote-Nete catchment with a higher precipitation level than Leipzig and Dessau, the slope of the runoff – degree of imperviousness curve seems rather comparable to the curves of Leipzig and Dessau.

The results for the declining pattern of infiltration between Leipzig and the other three studies differ strongly. According to the study effected in Leipzig until a degree of imperviousness between 50 and 70 percent no significant effect on infiltration decline can be seen. For Leipzig this seems to be the critical value for the degree of imperviousness. But Munich and Dessau show a linear or exponential decline of infiltration with an increasing degree of imperviousness and there is generally no critical level where infiltration is unaffected by surface sealing. So it seems that even the smallest degrees of imperviousness affect the infiltration which leads to the conclusion that surface sealing should be avoided as far as possible.

Even if the results show a clear tendency, the relationship of evapotranspiration and infiltration with degree of imperviousness is not uniform. Also the slope of the shown linear relationship between surface runoff and degree of imperviousness differs. This might reflect the reality due to different soil properties, different groundwater levels or deviating climate data. But also a share should be attributed to different modeling and definition of water balance components. For example, in Munich evapotranspiration was calculated as a residual from total precipitation after subtracting infiltration and surface runoff. Additionally, evapotranspiration in Munich includes the interflow as only the net infiltration is modeled. In Leipzig and Grote-Nete, the actual evapotranspiration was modeled by mainly considering the relationship with the potential evapotranspiration. Furthermore, in Leipzig and Grote-Nete climate data was used in gridded form, whereas for Dessau and Munich a unique value for the whole area was used. Also the matching of land-use type and degree of imperviousness can be sources of inaccuracy. We have to consider for example that in Leipzig the land use type detached houses is classified with a degree of imperviousness of 40 to 60 percent and in Munich the same type of houses has a degree of imperviousness of 30.5 percent (Pauleit & Duhme, 2000) (Haase & Nuissl, 2007). In Grote-Nete, urban land use was in general classified with a plug-in value of 70 percent without differentiation. Additionally, the size of the studied areas strongly differs from 30 km² in Dessau to 525.5 km² for the catchment Grote-Nete so that amount of underlying data has also influenced the quality of the results such as statistical outliers might have a larger effect in smaller studies. Additionally, the methodology in Dessau was insufficiently described to consider methodological differences to the other data and to assess the reliability of the data.

An improvement could be made by using more data points to reduce the impact of single measurement or modeling errors and to obtain more accurate estimates of the relationship between degree of imperviousness and the water balance components.

5. Conclusion

We have seen that with increasing degree of imperviousness, the water balance is increasingly altered. This has been shown for different levels of precipitation. Even minor alterations of land use impact significantly alter the water balance. The severe consequences of surface sealing on the water balance should be considered for all decisions on land use change.

Additionally, this relationship might be used to communicate more the severe consequences from urban sprawl to a broader public than by using hydrological properties including soil type or field capacity that are more difficult for laymen to understand.

As severe consequences for the urban water balance could be seen, it has to be thought more about how we can reduce the degree of imperviousness in cities. The major option is to reduce land consumption in form of urban sprawl. This is the further spreading of anthropogenic surface sealing outside the existing boundaries of our current settlements (Haase & Nuissl, 2007). Additionally, highly impervious materials used for roads or parking places such as tar or concrete could be substituted by semipervious lawn bricks or cobblestones (Nehls, Jozefaciuk, Sokołowska, & Wessolek, 2006). But it has to be considered that this is only a remediation with a smaller effect compared with the limitation of urban sprawl. Not all roads in a city can be replaced by semipervious materials for economic – as they are more expensive – as well as practical reasons as they cannot stand high traffic flows that concrete or tar can. Nevertheless, it could also be seen that roads in the study of Munich for example account for about 6 percent of the overall surface cover and housing and commercial area use are responsible for most land consumption. And in these areas with low traffic alternative pavements are suitable.

Further investigations could be made outside Germany, where most of the studies to investigate relationship of the degree of imperviousness and the water balance are effected to test whether the results of this report are applicable there (Scalenghe & Marsan, 2009). Additionally, it would be interesting to study how this relationship changes considering climate change, mainly the change of precipitation and temperature. Possible consequences could be a decrease of precipitation and increase in temperature in some European regions so that the water balance might be even more threatened. So the consideration of surface sealing in urban planning should become even more important (Gill, Handley, Ennos, & Pauleit, 2007).

In addition, it could be interesting to test how long it takes until the natural state of the water balance is disturbed – e.g. a decreasing groundwater table. From the other side, it could be interesting to investigate to which extent a reduction of surface sealing might heal an alteration of the water balance and how long it would take after a surface sealing reduction until the components will be in a nearly unaltered state again.

References

Asefa, T., Wang, Z., Batelaan, O., & De Smedt, F. (1999). Open integration of a spatial water balance model and GIS to study the response of a catchment. *Proceedings of the Nineteenth Annual American Geophysical Union Hydrology Days*, (pp. 11-22). Fort Collins, US.

Assouline, S., & Mualem, Y. (2002). Infiltration during soil sealing: The effect of areal heterogeneity of soil hydraulic properties. *Water Resources Research, 38* (12), pp. 221-229.

Bagrov, N. A. (1953). O srednem mnogoletnem isparenii s poverchnosti. *Meteorologiya i Gidrologiya, 10*, pp. 20-25.

Bhaduri, B., Minner, M., Tatalovich, S., & Harbo, J. (2001). Long-term hydrologic impact of urbanization: A tale of two models. *Journal of Water Resources Planning and Management, 127* (13), pp. 13-19.

De Smedt, F., & Liu, Y. (2004). *WetSpa extension, A GIS-based hydrologic model for flood prediction and watershed management.* Retrieved October 9, 2011, from http://www.vub.ac.be/WetSpa/downloads/WetSpa_manual.pdf

EEA. (2011). *EIONET Gemet Thesaurus - Soil surface sealing.* Retrieved September 29, 2011, from http://www.eionet.europa.eu/gemet/concept?cp=7897

EEA. (2011). *Land take (CSI 014) - Assessment.* Retrieved October 9, 2011, from http://www.eea.europa.eu/data-and-maps/indicators/land-take-2/assessment

EEA. (2010). *The European environment — state and outlook 2010 - land use.* Luxembourg: Publications Office of the European Union.

EEA. (2010). *The European environment - state and outlook 2010 - urban environment.* Luxembourg: Publications Office of the European Union.

Ellis, B. J. (2008). Third generation urban surface water drainage; From rooftop to the receiving water sub-catchment. *11th International Conference on Urban Drainage.* Edinburgh, UK.

Emmerling, C., & Udelhoven, T. (2002). Discriminating factors of the spatial variability of soil quality parameters. *Journal of Plant Nuitrition and Soil Science, 165* (6), pp. 706-712.

Gill, S., Handley, J., Ennos, A., & Pauleit, S. (2007). Adapting cities for climate change: The role of the green infrastructure. *Built Environment, 33* (1), pp. 115-133.

Göbel, P., Starke, P., Meßer, J., & Coldewey, W. G. (2011). A water sensitive city – Impacts on the urban water budget by the use of evaporation optimized pervious concrete pavements. *12th International Conference on Urban Drainage.* Porto Alegre, Brazil.

Google Inc. (2011). *Google maps - Surface sealing.* Retrieved October 4, 2011, from http://maps.google.com/maps/ms?msid=215371506603150073069.0004ae7ac460cc57ef872&msa=0&ll=49.037868,20.258789&spn=15.86052,39.506836

Haase, D. (2009). Effects of urbanisation on the water balance – A long-term trajectory. *Environmental Impact Assessment Review, 29*, pp. 211-219.

Haase, D., & Nuissl, H. (2007). Does urban sprawl drive changes in the water balance and policy? The case of Leipzig (Germany) 1870 - 2003. *Landscape and Urban Planning , 80* (1), pp. 1-13.

Hundecha, Y., & Bárdossy, A. (2004). Modeling of the effect of land use changes on the runoff generation of a river basin through parameter regionalization of a watershed model. *Journal of Hydrology , 292* (1-4), pp. 281-295.

Leopold, L. B. (1968). *Hydrology for urban land planning - A guidebook on the hydrologic effects of urban land use.* Washington D.C., US: U.S. Department of the Interior.

Mallants, D., & Feyen, J. (1990). *Kwantitatieve en kwalitatieve aspecten van oppervlakte en grondwaterstroming.* Leuven: Katholieke Universiteit Leuven.

Messer, J. (1997). *Auswirkungen der Urbanisierung auf die Grundwasserneubildung.* Essen: Deutsche Montan Technologie GmbH.

Nehls, T., Jozefaciuk, G., Sokołowska, Z. H., & Wessolek, G. (2006). Pore-system characteristics of pavement seam materials of urban sites. *Journal of Plant Nuitrition and Soil Science , 169*, pp. 16-24.

Pauleit, S., & Duhme, F. (2000). Assessing the environmental performance of land cover types for urban planning. *Landscape and Urban Planning , 52* (1), pp. 1-20.

Rasch, B., Friese, M., Hofmann, W., & Neumann, E. (2010). *Quantitative Methoden 1 - Einführung in die Statistik für Psychologen uns Sozialwissenschaftler* (3rd ed.). Berlin/Heidelberg, Germany: Springer Verlag.

Riemann, U. (1999). Impacts of urban growth on surface water and groundwater quality in the City of Dessau, Germany. In B. J. Ellis, *Impacts of urban growth on surface water and groundwater quality* (pp. 307-314). Wallingford: IAHS Press.

Rodhe, A., & Kilingtveit, Å. (1997). Catchment Hydrology. In O. M. Saether, & P. de Caritat, *Geochemical processes, weathering and groundwater recharge in catchments* (pp. 77-107). Rotterdam, Netherlands: A.A. Balkerna.

Scalenghe, R., & Marsan, F. A. (2009). The anthropogenic sealing of soils in urban areas. *Landscape and Urban Planning , 90*, pp. 1-10.